YOUR KNOWLEDGE HAS VALUE

- We will publish your bachelor's and
  master's thesis, essays and papers

- Your own eBook and book -
  sold worldwide in all relevant shops

- Earn money with each sale

Upload your text at www.GRIN.com
and publish for free

Bibliographic information published by the German National Library:

The German National Library lists this publication in the National Bibliography; detailed bibliographic data are available on the Internet at http://dnb.dnb.de .

Imprint:

Copyright © 2016 GRIN Verlag
Print and binding: Books on Demand GmbH, Norderstedt Germany
ISBN: 9783668775077

This book at GRIN:

https://www.grin.com/document/437224

Jean Cédric Obame Emane

Fractal Time by Gregg Braden. A Book Discussion and Overview

GRIN Verlag

JEAN CEDRIC OBAME EMANE

COURSE NAME: SEMINAR INTERNATIONAL DEVELOPMENT I – FRACTAL TIME – (ESSAY)

REGISTERED IN THE SCHOOL OF SOCIAL AND HUMAN STUDIES

STUDENT'S PROFILE:

INTERNATIONAL RELATIONS

MAJOR: INTERNATIONAL SECURITY

July 3rd, 2016.

A partial fulfillment of PhD degree in International Relations

Major: International Security

ATLANTIC INTERNATIONAL UNIVERSITY

Content

Cycles that repeat, the 20 year "curse "of American presidents and identification of the cycles of nature in our lives

I. Introduction

This book <u>Fractal Time</u> by Braden is fascinating. The author centers his analysis on the year 2012, which many people gave interpretation to. An interpretation that consisted of seeing that year as the end of the world. Besides, that way of viewing things for the year 2012 was a confusion of events. At least Braden wrote the book before the coming of that year, which proves that the explanations he gives in his book are trustful. When Braden works on the year 2012, his focus is about the cycles of times. Cycles of times that can occur in our day-to-day life and are more likely to repeat after many years. According to the author we have to approach this year the best way possible. To convey his message, he talks about 20 times codes that are developed in the book. Time code 1 says that we are living the completion of a 58,185-year-long cycle of time- a world age that the ancient Maya calculated would end the winter of the solstice on December 21, 2012. We are living the end of time, not the end of the world as a lot people predicted, but the end of a world age. For example, the present world age began in 3114 B.C. and will end in A.D. 2012. Time code 2 states that our ancestors recorded their experience of the last **end of time** showing beyond a reasonable doubt that the close of one world age is the beginning of the next, and not the end of the world. Now, the question might be: what is the meaning of the end of world age? Those who came before us knew the end of the time was coming, they knew because it always does. Accordingly, Braden asserts that every 5,125 years, the Earth and our solar system reach a place in their journey through the heavens that mark the end of such a time cycle. Only five generations in the last 26,000 years have experienced the shift of world ages. We will be the sixth. The only way to arrive at the light of the next cycle is to finish the darkness of this one. Many themes are developed in Braden's book. I have selected those I thought to be more appealing. The first issue I will discuss are cycles (time cycles) that repeat, and then I shall debate the 20-year curse of American presidents. The last discussion will be the identification of cycles of nature in our lives.

II. Cycles that Repeat

Somewhat, it might be unbelievable that cycles exist and can repeat in our lives or whatsoever. Some critics might attribute them to coincidence, but not as something that had already occurred in times. To explore that issue, we will see what Braden (2009) says about those patterns. Braden (2009, chapter 5) starts with a shocking story, the assassination of President Kennedy in 1963. The author experienced that event when he was a child. On that sad day, his teacher told them to remain calm before collecting their coats and suppliers because school buses were waiting for children outside of the school. It was too early to go home, that was the middle of the day. Braden asked his teacher what was going on, she replied that they had no right to talk about that issue with children. Parents will. When he arrived home, he found his mother with tears and who finally told her to watch the television, solely to see that the American president had just died.

Two days later, the local media printed a story that actually drew Braden's attention. The story was about the curious circumstances about his assassination. The title of the story was "History Repeats Itself". The main point was the strange coincidences that related the 1963 assassination of President Kennedy to another that had happened about 100 years prior to that- that of President Abraham Lincoln. Both presidents were deeply implicated in racial equality and civil rights. The two presidents had wives who lost children while being in office. The two presidents had been murdered on a Friday. Both had died from a gunfire wound to the head.

Braden, at the time he was reading the article simply did not want to take it seriously. Lincoln for instance, was sitting in box 7 of Ford's Theatre when he was shot. Kennedy was riding in car number 7 when he was killed. The two presidents were with their wives at the time they were shot. Before being president, Lincoln entered elective office in 1846. A century later, in 1946, Kennedy was elected to Congress. Lincoln became president in 1860, Kennedy in 1960. The last name of the men who replaced them in office was identical: Johnson. And both Johnson were born 100 years apart. President Andrew Johnson was born in 1808; meanwhile Lyndon Johnson was born in 1908. As the article kept moving with those comparisons, to Braden they seemed to be more than coincidences, in so much as they went further than the mere assassination of both presidents. Both presidents had four children and both had lost two of them before their teenage age. Both had lost their son while in office. The name of both presidents 'doctors was the same: Charles Taft. The name of Lincoln's private secretary was John (the first name of Kennedy), and that of Kennedy's secretary was Lincoln (the last name of President Abraham). In addition, the events even extended to their murderers. The name of the man who arrested Lincoln's shooter was Baker. The name of the man who arrested Kennedy's assassin was also Baker. According to Braden, it appeared that those patterns were endless; he as well recognized that they were irrefutable. It was difficult to understand why and how those two events with a 100-year interval could be so similar. The truth is that they were. Although we may not want to acknowledge it or not, history repeats itself in the case of both presidents if we take into consideration these two occurrences.

I have to admit that Braden's books are strange in a sense that as in the case of The Divine Matrix, you always discover things that you did not know or could not explain before reading the book. How is it that there are some cycles in our lives or other people's lives or in our environment? How could things that occurred in the past come to pass again? Maybe those phenomena might be understandable in the example of catastrophism – the idea that many of Earth's crucial features (strata layers, erosion, polystrate fossile, etc.) formed as a result of past cataclysmic activity – that happened in the past, I mean things that occur naturally. However, when it comes to things that occur to people, we start having headaches. When I try to figure out what can explain the events that happened to both presidents, I think about the Matrix Braden (2007) talks about. It is the Matrix that has the power to cause events that once occurred in the past to repeat.

What I would like to point out is that for those things to repeat a supernatural intervention is required. That supernatural intervention can only be made possible through the Divine Matrix. I refer to the Matrix because Braden (2007, p. 25) gives us an explanation to understand what had happened to both presidents. He states that everything in our world is connected to everything else (key 2). For the author, we are all connected one another, connection exists between us and the world around us. What I mean is that both presidents were already connected although they lived in different times; they were connected spiritually in times. I believe the author is right when he says that connection exists between everything within the Matrix. Remember the experience I shared in my previous book analysis in Braden's Divine

Matrix's book about being connected to things that had not happened yet. With this in mind, Braden (2007, p. 27) supports that connection within the Matrix goes far beyond what we see in our lives today, it has to do as well with everything we have ever seen and also with things that have not materialized yet.

This is the experience I shared. Many, many, many times I might have found myself with some friends drinking, or eating or simply talking at a friend's place, or in a bar or whatever, and had the feeling that I had already been to that place. What I mean is that the kinds of conversations we shared at that very moment seem to be the conversations I had already shared with them while I had never been to that place neither my friends. At the time I would experience such a thing, I would try to find out what was happening. Because I had the firm belief that me and my friends had already been there (physically) and been witnesses of the event, but in reality we never went to that place before. The reason is that my mind was already connected to that event and my friends prior to the occurrence of that situation.

I would like to point out that Presidents Lincoln and Kennedy were connected, that is the reason why they might suffer the same karma. We cannot deny that something out of the ordinary connected these people. Accordingly, they undergo the same fate. We can notice that the cycles of darkness repeat in their lives dangerously. I have just discussed the part about cycles that repeat, now I would like to debate the 20-year curse of American presidents.

III. The 20-year Curse of American Presidents

Braden (2009, chapter 5), declared that a curse on American presidents might have existed, that is the fact that every 20 years a US president dies. The author stated that historians worked on the issue and started to suspect that the assassination of President Lincoln in 1863 was the beginning of a cycle that started a little more than 20 years earlier when another US president died. William Henry Harrison had become sick and died of pneumonia. With his death it looked as if the seed (cycle) had been engrained from that very moment, to pattern such tragic events. In the years after those events, the suspicions of those historians proved to be truthfully real.

As strange as it may seem, in the nearly 160 years after the death of President Harrison, almost every 20 years the American president has either died in the White House or has survived a shot on his life. As an illustration we have the unsuccessful attempts to kill Ronald Reagan and George W. Bush. Braden figures out that scholars question if these attempts to kill the cited presidents have put an end to the 20-year curse of American presidents. Braden says that the presidential cycle of 2020 will be the answer. But when observing the statistics, they appear to be obvious. The following table in the book of Braden will illustrate better the so called 20-year curse.

Year Elected	President	Event
1840	William Henry Harrison	Died in office
1860	Abraham Lincoln	Assassinated
1880	James Garfield	Assassinated
1900	William McKinley	Assassinated
1920	Warren Harding	Died in office
1940	Franklin Roosevelt	Died in office
1960	John F. Kennedy	Assassinated
1980	Ronald	(Survived assassination attempt)*
2000	George W. Bush	(Survived assassination attempt)**

- Bush avoided injury from a grenade that was tossed in his direction during a 2005 visit to Georgia, the former Soviet satellite.

This table shows that since the election of the United States President in 1840, the country has lost sitting president to illness or violence every 20 years.

The table obviously reveals that the 20-year curse seemed to have been a reality. Here again comes the question related to cycles. How is it that statistics prove that since 1840, a US president has either died in office or been assassinated? It is a cycle that I believe we can absolutely do nothing about. According to the question Braden asked whether or not the seed or the cycle stopped since the failure to shoot Reagan and Bush occurred, I believe that nobody has the answer. As the author pointed out, the year 2020 presidential elections will determine the answer to that question. Because it will culminate 40 years if no president is assassinated.

To conclude this part of my dissertation, I shall agree with Braden when he says that whether we talk about the cycles that repeated between President Kennedy and President Lincoln, or the 20-year curse American presidents witnessed, three things are palpable:

Fact 1: There are cycles underlying both events.
Fact 2: Both cycles are "triggered" by a seed event.
Fact 3: The conditions of the seed event repeat at regular intervals.

This apparently illustrates time code 15 that states that patterns identified for an earlier time in history tend to repeat themselves with greater intensity at later dates.

I have just debated the 20-year curse US presidents have faced. Here and now, let me tackle the last subject of my book review: identification of the cycles of nature in our lives.

IV. Identification of the Cycles of Nature in our Lives

Regarding that subject, Braden (2009, chapter 5) provides the definition of the cycles of nature. The cycles of nature are referred to as our personal lives as well as to global events. He sustains that although we might be aware of the relationship in the cycles of nature, those cycles happen to us when we do not expect them. The Braden shared the first time the cycles of nature entered his life. He talks about the seed of love and betrayal. He was eleven when he first experienced the cycle of nature.

A Saturday morning when waking up and was still in his bedroom he felt like the atmosphere at home was different, he could sense that something was going to happen. His mother was in the kitchen and he could perceive that she had been crying. As the sentiment he had was so striking, he decided to tiptoe to his parents' bedroom only to discover his father putting his cloths on his suitcase. As his father saw him from the corner of his eyes, he called his son to come near him. He told Braden that he loved him, but that he was going to leave home for sometimes and that he did not know when he would be back. He and his mother observed his father leaving the house. At that time Braden did not know that he had just witnessed the divorce of his parents.

It is only after some years that he realized that this particular event actually affected his life. He noticed that he did not only lose his father but also his family; he meant the way the same was until his 11 years of life. He thought that his life would continue to be the same, but soon realized that something was missing in it, that on that Saturday morning he lost something else. He shared that experience to explain that there may be events in our lives that can be the seed event as seen above, that can foster them to repeat in other circumstances. Braden asserts that cycles that might occur in our lives are not a problem if they show positive patterns. The issue is when these patterns are negative.

Luckily, just as repeating cycles are as well chances to change the patterns for conflict and violence on a global scale, if we know our individual cycles and how they work, they could constitute influential partners in healing some of the highest pains of our lives. Braden says that the time code calculator is a good tool to find out such cycles in our lives. In that perspective, time code 13 is particularly interesting when it states that our knowledge of repeating cycles allows us to pinpoint times in the future when we can expect to see the repeating of the past. In order to know about when future events might repeat in our life, Braden proposes the time code calculator. Time code 14 for example says that the time code calculator can pinpoint personal cycles of loves and hurt, as well as global cycles of war and peace. The point is that even if his mother, his younger brother and him pretty much looked like a family, Braden felt the loss of his family. He had the sentiment of betrayal and loss.

Braden (2009, chapter 5) has decided to use the time code calculator to know when the event when the seed was first planted is more likely to occur again. He uses his experience of loss and betrayal to illustrate that point. To perform it he suggests three steps:

- Step 1: identify your age at the time of the seed. In his case he was 11 years old.
- Step 2: calculate the phi ratio interval of your age during the seed event. In his case: 618x11= 6,798.
- Step 3: add the phi ratio interval back to your age during the seed event to determine your age when the conditions will repeat. Example: 6798x11= 17,798.

From this simple calculation it seems that the age at which the sad experience of his 11 years old is more likely to happen again is 17,798.

Braden concludes that the cycles of our lives are not to be considered as to be absolute events. The reason for it is that we are talking about normal processes that respect the rhythms of natural cycles. As for him, our choice is accountable for changing the course of a given cycle. The secret in exploring personal cycles is firstly to admit that they are happening; secondly the frequency with which they repeat.

When considering Braden's statement we might notice that he witnessed his first seed event (cycle) at the age of 11 years old, a moment that particularly affected his personality. Because he recognizes that, although his mother, little brother and him looked like a family, his sense was that he lost something more than his father the very day he left home that Saturday morning. He has a sentiment of loss and betrayal. That seed event is more likely to be repeated in the future. The author suggests something very interesting, that is, we have to be able to identify cycles of natures. Equally important, they can help prepare for the next step in the cycle. Braden explicates that when the patterns that come with the seed event are positive, we seem not to remark that a seed has been planted. It is rather when the seed has negative patterns that we can lose our smile.

I believe using time code calculator is interesting in a sense that it might have us know when the next seed event (cycle) may occur again. But still, I do not give credit to the time code calculator, only in one circumstance. Let us examine his time code calculator that gave the age 17,798 as the age at which what happened to him at 11 might repeat. That calculation might give an idea of when the event might come to repeat, however, in the case of Braden, is there a human being who ever lived until the age 17,798?

It means that when the seed event repeats, Braden would have disappeared from the earth a long, long, longtime ago in so much as it is not documented that a human being has reached that age. I said that I can give credit to time code calculator only in one circumstance. Accordingly, if something ever happens to me, the time code calculator might help share it with my descendants, that they share to their descendants when that cycle might occur again. That is the only option for the reasons that I have already provided in the case of Braden.

When Braden says that we have to be aware of natural cycles that may enter our lives and recognize that they are actually happening; and when he states that the choices that we make at any time in our lives can forever change the course of a given cycle, there is a problem. Does it mean for instance that he has the power to prevent his 11 years old experience from happening again because of a specific choice he could have made? This part of the book's review has been very edifying as it showcased how natural cycles 'seed can enter somebody's life (the case of Braden at his 11 years old).

Let us examine an article by Mary Darling Montero, *Is History Repeating Itself in Your Relationships*? According to her, a number of times people came to see her for a therapy because they had the impression that there were some patterns that were repeating in their relationships. She asserts that the majority of people unconsciously often follow diverse patterns in their relationships – patterns that might have been planted in their personalities. Nevertheless, if they take a look at themselves, they will soon notice that they carry with them old patterns that can affect their present relationships. The key is to be aware of their existence (as Braden suggests) and learn how to change them. The author says that if a patient goes to see her after a breakup or a divorce, she would ask them to examine not only the relationship in question, but also all other significant relationships the person has ever had in their life, maybe by turning back to childhood, the relationship with their caregivers, parents. Because patterns that were planted in that period might repeat.

She points out that the well-known belief is that these patterns are largely unconscious, working in areas of the brain that are not often connected to our consciousness. The idea that the way our parents or caregivers treated us or the kind of relationship we had with them governs unconscious patterns of security, trust and independence is well documented. Understanding that history repeats itself can be viewed within two possible frameworks: patterns that are persistent all along our lives, and an isolated event that conveys unexpected feelings or behaviors.

As an illustration, she reports the case of her patient "Jaclyn" who discovered after therapy that throughout her professional, romantic, social and familial relationships she had the tendency to put her own needs backward, that the author qualified as a "people pleaser". Consequently, her own needs were not being considered on a constant basis. She figured out that the explanation was to be found in her childhood. Her parents divorced and she felt neglected and she thought that the only way to get their attention and approval was by pleasing them. This character influenced her adulthood in a positive way given that assuming herself in healthy ways did not cause people to leave her. Rather it generated better relationships and gave her more confidence in herself.

With regard to the second framework, that of isolated incidents, the author reports another case of a patient name "Paul". When Paul's wife got a promotion at work she began to work longer hours, which means she became less and less available for her husband (both physically and emotionally). Then Paul began to stay away from his wife. He became passive-aggressive by responding to his wife with mockery and what he himself called "dirty look". He was annoyed with the kind of behavior he had toward his wife and said that he had never acted that way before in his life. However, he figured out that he definitely acted like that toward his mother when he was at primary school. His mother had got a job after being a stay-at-home mother since his birth. As he was too young to know his feelings, his reaction was to ignore his mother and exaggerating his anger. He realized that this memory came back to his mind and acted with his wife exactly the same way he did with his mother, as a child. After that flashback to his past, he was capable of coping with his wife's new situation.

The author as Braden, sustains that she shared these glaring cases to evidence that although these cycles might exist in our life, there are ways to alter them, solely that we have to go down to our memories in the seed event of our past relationships.

This article shows the obviousness of cycles of nature in our lives. Things whose seed events as in Braden's case started in childhood tend to repeat in our adulthood at the time we do not suspect them to happen in our lives. The author gives us clues to understand repetitive patterns in our relationships, which are things that always occur in our lives exactly the same manner. She suggests that we examine the relationships we have, but also the kind of relationships we had in our childhood. In effect, the key is mostly located in the patterns that had taken place in our childhood.

As a result, these patterns can affect our lives in a positive or negative way depending on their nature. In the case of Jaclyn, she had a positive pattern that turned to repeat as a cycle in her life because of her parents' divorce in attempting to act in a way that pleased them at the time. The patterns helped her because it caused her to make many friends and gave her confidence in herself. In the case of negative patterns, she reports the story of Paul. Obviously, the cycles that repeated in the life of Paul proved to influence his life in a dangerous way, in a sense that they could have jeopardized his marriage.

In childhood, the pattern had no negative influence since his mother might have considered it as a caprice. Meanwhile in adulthood that would probably put his marriage in danger in so much as his wife could not have tolerated his attitude as described above. Montero wants to point out that there is a good news: we can change our pattern and prevent them from jeopardizing our adulthood. According to her, we have to look back to our memories in order to understand the patterns occurring in our relationships. We can notice that the technique seems to bear fruits if we take into account the case studies she provided in her article.

V. Conclusion

This book <u>Fractal Time</u> by Gregg Braden is actually great in a sense that it helps the reader realize that patterns exist in the human condition. Braden shows that it is impossible not to take that issue seriously because if we cautiously observe our lives, we shall realize that these cycles really exist. Braden recommends that we recognize that these patterns exist, and then calculate when they are more likely to happen again. In my essay, I have debated three specific subjects: cycles that repeat, the 20-year "curse" of American presidents and the identification of the cycles of nature in our lives. On the first topic, we have seen that the assassination of President Kennedy in 1963 appeared to be a pattern that occurred about one hundred years earlier. It is linked with the assassination of President Lincoln of 1860 in so much as all the patterns that had happened in the life of the latter repeated in the life of the first one. As for the 20-year curse of American presidents, we have seen that from 1840, the seed event of US president's assassination was planted. Indeed, from that year statistics revealed that every twenty years, an American president dies in office, is assassinated or survived deadly attacks. Concerning the last theme of our dissertation, we have seen that seed events can be planted in our lives and can constitute the beginning of the cycles of nature. <u>Fractal Time</u> is remarkable as book in that it shows how the Maya calculated the end of times corresponding with the year 2012. A lot of people were panicked because of that year. But ahead of time, Braden did a great job in demonstrating that the year 2012 had nothing to do with the end of the world, but the end of the world age.

References

Braden, Gregg (2007). <u>The Divine Matrix: Bridging Time, Space, Miracles, and Beliefs.</u> Australia: Hay House.

Braden, Gregg (2009). <u>Fractal Time.</u> Australia: Hay House.

Montero, Mary Darling. *Is History Repeating Itself in Your Relationships?* Retrieved from: http;//m.huffpost.com/us/entry/672518.html.